A Horizontal Langstroth Bee Hive

You CAN build this!

How to Build a Horizontal Langstroth Bee Hive
Copyright 2020
New Hampshire WoodWorks
Northwood, New Hampshire

How to Build a Horizontal Langstroth Beehive

By W. Todd Abernathy

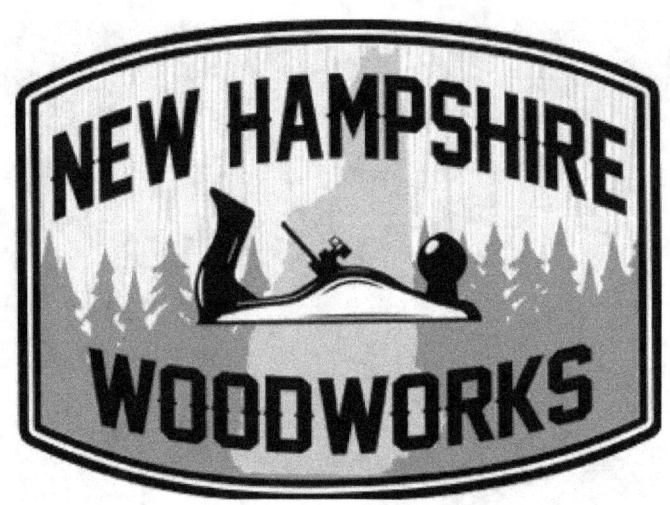

A Homesteading 'How To' Book From

www.UnexplainedUnderfootObjects.com
&
www.NewHampshireWoodWorks.com

Other Books in the Homesteading 'How To' Series:

The Langstroth beehive is a standard design for many beekeepers the world over, but it isn't the only choice beekeepers have. In this 'Homesteading How To' beehive building series, we share a few that we've build and modified over the years, each with their own strengths and characteristics.

The Horizontal Langstroth is designed for ease of access while still using traditional Langstroth frames. The design incorporates traditional bee space in a single, stationary, and waist-level unit without having to lift and stack 50-pound boxes.

The Horizontal Langstroth is great for single hive apiaries and decorative garden placement. It is wheelchair friendly and the gabled roof is hinged for ease of access.

In this 'Homesteading How To' book, you will learn how to build a simple and well-designed Horizontal Langstroth beehive that will serve you and your bees for years to come. The materials are inexpensive, the skills required are novice-level, and in the end, you will have something beautiful that your bees will flourish in.

List of Tools

- Simple squaring jig (plans on the next page)
- Table saw with edge guide fence
- Miter saw
- Handheld drill (2 for efficiency)
- 5/64" drill bit (countersink optional)
- (4) 24" F-bar clamps
- Jig saw

Lumber for Squaring Jig

- 4ft x 1" x 2" oak board
- 4ft x 1" x 3ft sheet of plywood

Lumber for a 4' Horizontal Langstroth

- (4) 12ft x ¾" x 12" select pine boards
- ¾" exterior plywood for roof
- (4) 36" lengths of 2x4" pine for legs

Hardware

- (1) box of #8 1-5/8" exterior deck screws
- Bolts/washers/nuts for legs
- Front Handle
- Chain lid support
- 2 Hinges
- Quality wood glue

Squaring Jig Plans

On one long side of the 4ft x 1" x 3ft plywood, match the strip of 4ft x 1" x 2" oak board so it overlaps the plywood top by ¼". (You can rout out the oak to snug-fit the plywood for even more support before fastening, but it's not necessary.)

Pre-drill screw holes at intervals to avoid splitting the plywood. Fasten with at least 6 evenly-spaced decking screws to keep the oak secure and square at 90 degrees to the plywood.

The squaring jig will sit on top of any table saw (with its blade retracted) and help as a condensed work surface. If the cuts are square, the boxes will square themselves naturally against the edge during fastening.

To use, stand one long side of a bee box against the fence and build out from there.

Master Measurements (using 3/4" Pine)

Hive Box:

2 pieces 48" x 8 15/16" (long sides)

2 pieces 17 13/16" x 8 15/16" (short ends)

1 piece of ¾" Plywood or equivalent pine boards 48"x19 3/8"

2 pieces for landing boards 10" x 2 ½"

2 pieces 5" x 2 ½" for Entrance Reducers

2 pieces 48" x 4.5" side wrap boards

2 pieces 20 7/8" x 4.5" end wrap boards

2 pieces lid support 49 ½" x 4.5"

Interior cover boards 19 1/4" length, width determined by need

Base Stand/Entrance:

4 pieces of 36" 2"x4" for legs

Roof:

2 pieces 49 1/5" x 4 ½" (sides)

2 pieces plywood or pine equivalent 51" X 14"

1 piece 51" x ¾" x ¾" roof peak

3 pieces gabled support as shown:

Build the Lower Box

Hive Box:

2 pieces 48" x 8 15/16" (long sides)

2 pieces 17 13/16" x 8 15/16" (short ends)

1 piece of ¾" Plywood or equivalent pine boards 48" x 19 3/8"

2 pieces for landing boards 10" x 2 ½"

2 pieces 5" x 2 ½" for Entrance Reducers

2 pieces 48" x 4.5" side wrap boards

2 pieces 20 7/8" x 4.5" end wrap boards

2 pieces lid support 49 ½" x 4.5"

Before you fasten your sides, it's recommended to make a dry fit with the box sides and a Langstroth frame to ensure the proper bee space exists between the frame edge and the sides. The bee space should be roughly 1/8" of an inch on each side. If you don't have enough in your dry fit, extend your end piece measurement accordingly, as well as your bottom board widths, recut, and telegraph that measurement in future end cuts.

Using the squaring jig and clamps, bring together the four sides of the box. Predrill four holes at each junction with a counter sink drill bit and fasten with your deck screws.

As a quick note, during the build I kept a few Langstroth frames on hand to double check depth and width, and this practice helped me find errors as I progressed. I highly recommend this to save time and money.

Now, attach your bottom board(s). This will either be the plywood (48" x 19 3/8") or lengths of ¾" pine. We used pine in this build, but either material is fine for our purpose.

Mark your entrance holes on the bottom of your box bottom and use a jigsaw to remove the waste. We cut ours at 12"x6". This allows for a 5" entrance reducer made from ¾" scrap.

Once the entrance is cut, attach a landing board measuring 10" x 3" over the bottom hole with a lip to act as a landing board. As shown, we placed two entrances with two landing boards.

Once the bottom board is all set, flip the box to finish the lower build.

Attach your wrap boards (48" x 4 ½" sides and 20 7/8" x 4 ½" ends) with a 1 ½" gap above the existing box boards to create a lip for the frames to rest on as well as the inner lid boards. If you recut to adjust for bee space, take this into account on your wrapped ends. Simply cut to fit.

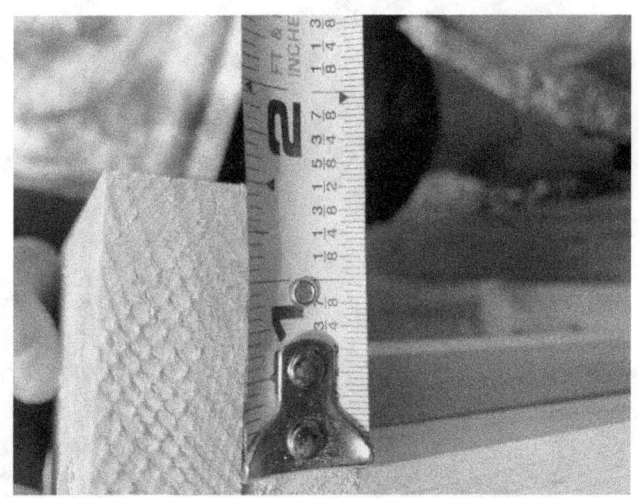

Lastly, before moving onto the lid build, cut your inner-lid pieces that cover the actual Langstroth frames. These will be 19 ¼" long, and as wide as you want. Because of the tight fit, we attached a small knob to our center board to help pull it out when working in the hive.

Build the Gabled Lid

3 gabled supports cut as shown below

2 pieces 49 1/5" x 4 ½" (sides)

2 pieces plywood or pine equivalent 51" X 14"

1 piece 51" x ¾" x ¾" roof peak

Cut three gable supports at 20 ¾" in length, with the center height at 9", and the sides at 3 ¼" as shown.

Connect these, with one on each end and the third in the center, with your 2 pieces of 49 1/5" x 4 ½ sides (flush with the bottom). Fasten on your 51"x14" roofing sides, then center your roof peak trim and screw into the gap created by the roofing sides.

Place finished roof on top of your lower box, attach your hinges, front handle, and chain hinge.

As shown, we have not attached any legs. However, you may easily fit 2"x4" legs cut to a desired length for your own stand.

All you need to do now is place your Langstroth frames in, protect the exterior with a stain, paint or spar urethane and invite some bees to set up housekeeping.

As you gain confidence with building bee hives, you will discover slight adjustments in measurements, hardware and materials that can make the build more efficient and productive for your particular situation.

Just remember the basics, and have fun!

Best Wishes,

W. Todd Abernathy
New Hampshire WoodWorks

NOTES